Rihab BEN ABDALLAH KOLSI

Renal complications in diabetics

Rihab BEN ABDALLAH KOLSI

Renal complications in diabetics

Diabetes and its complications

ScienciaScripts

Imprint

Cover image: www.ingimage.com

This book is a translation from the original published under ISBN 978-3-8416-2371-3.

Publisher:
Sciencia Scripts
is a trademark of
Dodo Books Indian Ocean Ltd. and OmniScriptum S.R.L publishing group

120 High Road, East Finchley, London, N2 9ED, United Kingdom
Str. Armeneasca 28/1, office 1, Chisinau MD-2012, Republic of Moldova, Europe
Managing Directors: Ieva Konstantinova, Victoria Ursu
info@omniscriptum.com

Printed at: see last page
ISBN: 978-620-8-56227-4

RENAL COMPLICATIONS IN DIABETICS

<u>*Rihab Ben Abdallah Kolsi*</u>

Faculty of Science, Sfax, Tunisia

E-mail: rihab_b86@hotmail.com

Contents

Introduction

Introduction

Today, diabetes mellitus is a major public health problem. Type 2 diabetes is the most common form, accounting for around 90% of all forms of diabetes. The ever-increasing frequency of this condition may be explained by lifestyle changes, particularly sedentary lifestyles, high-fat diets, obesity and an ageing population. The seriousness of this condition is linked to the occurrence of metabolic, infectious and degenerative complications, including renal manifestations.

Diabetic nephropathy (DN) is one of the most frequent complications of diabetes, progressing to chronic renal failure. Early diagnosis of this pathology enables patients to be managed more effectively and in a multidisciplinary way, delaying progression to chronic renal failure by inhibiting the renin angiotensin system (RAS) through good glycemic control, preventing or limiting iatrogenicity, and treating any complications.

This is the context of our work, which involved a prospective study of diabetic patients with or without ND with the aim of :

- ✓ Determine the distribution of diabetes according to gender, age and renal complications.
- ✓ Study selected biochemical parameters of blood glucose and renal function (blood glucose, HBA1c, urea, uric acid and creatinine) in people with diabetes and/or ND.

For this purpose, we followed a population of patients suffering from diabetes and/or diabetic nephropathy. The sociodemographic and clinical data of the patients were collected from a file described by the physician in order to select the subjects to be included in our study according to previously established criteria. This work was followed by research and analysis of various physiological and biochemical parameters in medical analysis laboratories.

Literature review

A. Diabetes

1. Definition

Diabetes is a serious chronic disease that occurs when the pancreas does not produce enough insulin (the hormone that regulates blood sugar levels), or when the body is unable to effectively use the insulin it does produce. Hyperglycemia, a common consequence of uncontrolled diabetes, can, over time, lead to serious heart, vascular, eye, kidney and nerve damage (WHO; 2016).

2. Epidemiology

2.1. In the world

Worldwide, an estimated 422 million adults were living with diabetes in 2014, compared with 108 million in 1980. The global prevalence of diabetes has almost doubled since 1980, from 4.7% to 8.5% in the adult population. These figures indicate an increase in associated risk factors such as overweight or obesity. Over the past 10 years, diabetes prevalence has increased more rapidly in low- and middle-income countries than in high-income countries (WHO; 2016).

2.2. In Tunisia

Diabetes is gaining ground in Tunisia. The incidence of diabetes rose from 13% to 15% between 1997 and 2005, and will rise to 27% by 2027. The number of diabetics is currently estimated at around 1.1 million, the majority of whom are undiagnosed, According to estimates this number will double in Tunisia in ten years to almost two million diabetics. (Abid M ; 2017)

3. Diabetes classification

WHO (2016) distinguished two main types of diabetes: insulin-dependent diabetes (IDDM) and non-insulin-dependent diabetes (NIDDM) although other types, can be included. These include gestational diabetes, diabetes related to malnutrition, glucose intolerance, etc.

3.1. Type 1 diabetes (T1DM)

Type 1 diabetes (10-15%) (previously known as insulin-dependent or juvenile diabetes) is caused by an autoimmune reaction in which the body's own defenses attack the insulin-producing beta cells of the pancreas. The body is then unable to produce the insulin it needs. The disease can affect people of any age, but usually appears in children or young adults (Nam H.-C et *al*; 2013).

3.2. Type 2 diabetes (T2D)

Type 2 diabetes (85-90%) (previously known as non-insulin-dependent or mature-onset diabetes) is a metabolic disease characterized by chronic hyperglycemia, the pathophysiological features of which include increased resistance of peripheral tissues (liver, and adipose tissue) to the action of insulin (insulin resistance), insufficient insulin secretion by pancreatic β-cells, inappropriate glucagon secretion, and a diminished effect of incretins, intestinal hormones that stimulate postprandial insulin secretion. (Chaudhry A et *al* ; 2013)

3.3. Gestational diabetes (GDM

Gestational diabetes (14%) is a temporary condition that occurs during pregnancy and is associated with a long-term risk of T2DM. The condition is present when blood glucose levels are higher than normal but still below the thresholds set for the diagnosis of diabetes. (WHO; 2016)

4. Diabetes symptoms

4.1. Symptoms of type 1 diabetes

Diabetes does not always manifest itself in the same way, with the same intensity and with all these symptoms. (Diabète Québec; 2014) In fact, T1DM appears suddenly and results in symptoms such as:

- Excessive thirst and dry mouth (polydipsia).
- Abundant and frequent urination (polyuria).
- Frequent urination.
- Lack of energy, extreme fatigue.
- Constant hunger.
- Sudden weight loss (emaciation).
- Slow wound healing.
- Recurrent infections..... (Nam H.-C. et *al* ; 2013)

4.2. Symptoms of type 2 diabetes

Symptoms may be similar to those of T1DM, but are often less marked or absent. (WHO; 2016)

- Excessive thirst.
- Profuse urination.
- Fatigue.
- Slow wound healing.
- Recurrent infections.
- Tingling or numbness in hands and feet..... (WHO ; 2016)

4.3. Gestational diabetes symptoms

In most cases, the mother has no symptoms, i.e. no typical signs of diabetes (strong sensation of thirst, frequent urination, for example). In many cases, it is non-specific symptoms such as increased susceptibility to urinary tract infections, high blood pressure, excessive amniotic fluid or high sugar excretion in the urine that point to diabetes (Swiss Diabetes Association).

5. Diabetes risk factors

5.1. Risk factors for type 1 diabetes

The causes of this destructive process are not fully understood, but a genetic susceptibility combined with environmental triggers, such as viral infection, toxins or certain dietary factors, is implicated. The disease can develop at any age, but type 1 most often appears in childhood or adolescence. (Nam H.-C et *al* ; 2017)

5.2. Risk factors for type 2 diabetes

Overweight and obesity, along with a sedentary lifestyle, are considered to be responsible for the largest share of the burden of diabetes-related disease worldwide (WHO; 2016)
Several dietary practices are linked to unhealthy body weight and/or T2D risk, including high saturated fatty acid intake, high total fat intake and insufficient dietary fiber intake.

In addition, active smoking (as opposed to passive smoking) increases the risk of type 2 diabetes. The risk remains high around ten years after smoking cessation, regressing more rapidly in light smokers. (WHO; 2016)

5.3. Risk factors for gestational diabetes

GDM appears following a decrease in insulin action (insulin resistance) due to hormone production by the placenta (Nam H.-C et *al*; 2017). Other risk factors for gestational diabetes include age, overweight and obesity, excessive weight gain during pregnancy, family history of diabetes, gestational diabetes in a previous pregnancy history of stillbirth or birth of a newborn with congenital anomalies and abnormal presence of glucose in urine during pregnancy (WHO; 2016).

6. Diagnosis of diabetes

Current WHO criteria advocate observing elevated blood glucose levels to diagnose diabetes. (Nam H.-C et *al*; 2017)

6.1. Blood glucose

The diagnosis of diabetes is based on the measurement of glycemia (blood sugar level) and the patient will be considered diabetic in the following situations: (Nam H.-C et *al*; 2017)

- Fasting blood glucose (no calorie intake for at least 8 hours) greater than or equal to 1.26 g/ l or 7mmol/ l.
- Blood glucose at any time of the day in the presence of clinical signs of hyperglycemia
  clinical signs of hyperglycemia (polyuria, polydipsia, unexplained weight loss often associated with polyphagia) greater than or equal to 2 g/l or 11.1 mmol/ l.
- Blood glucose level at 2th hour of OGTT (orally induced hyperglycemia according to WHO recommendations using an oral load of anhydrous glucose equal to 75g dissolved in water) greater than or equal to 2 g/ l or 11.1 mmol/ l.

Normal blood glucose values are less than 1g/ l fasting and less than 1.40 g/ l at the second hour of an OGTT.

6.2. Glycated or glycosylated hemoglobin (HBA1c)

For several decades, the diagnostic criteria for diabetes have been based on glycemic values such as fasting or postglucose blood glucose. In 2010, the American Diabetes Association (ADA) approved the use of glycated hemoglobin as a diagnostic tool for diabetes and prediabetes. Threshold values for HBA1c were established for diabetes ≥ 6.5% and for prediabetes between 5.7 and 6.4%. (K. Gariani et *al* ; 2011)

The value of HBA1c as a retrospective marker of glycemic control in diabetic patients is well established. HBA1c is a cumulative reflection of average blood glucose levels over the four to six weeks (up to three months) preceding the assay, and is used in current practice to retrospectively assess treatment efficacy. The risk of chronic complications, both micro- and macroangiopathic, is closely associated with HBA1c values (K. Gariani et *al*; 2011).

B. Diabetic nephropathy

1. Definition

Diabetic nephropathy (DN) is the most serious microangiopathic complication of diabetes. It involves glomerular sclerosis and fibrosis induced by the metabolic and hemodynamic changes brought about by diabetes mellitus. It manifests as slowly progressive albuminuria with worsening hypertension and renal failure. (Navin Jaipaul; 2018)

2. Symptoms of diabetic nephropathy

In the early stages of diabetic nephropathy, diabetics may not yet feel anything (no pain, no visible change in urine output). The following symptoms are likely to occur:

- Accumulation of fluid in the legs.
- Fatigue, exhaustion.
- Dyspnea (difficulty breathing).

Late signs may include weight gain and swelling of the ankles (edema). Nausea, vomiting, lack of appetite, weakness, itching, muscle cramps and anemia may also occur.

3. Diagnosis of diabetic nephropathy

According to the recommendations of the French National Authority for Health, diabetic nephropathy should be diagnosed by :

- **Testing for proteinuria** using urine reagent strips. This test should precede any test for microalbuminuria. The urine dipstick is more sensitive to albumin than to other proteins. The result may be :
  - Normal or insignificant: negative, traces, 1+ (corresponds to approx. 0.3 g/l).
  - Abnormal: 2+ (corresponds to about 1 g/l), 3+ (corresponds to about 3 g/l) or 4+ (corresponds to about 5 g/l) (HAS; 2014)
- **Microalbuminuria is measured** on a urine sample, particularly the first micturition in the morning**.** If this is positive, a 24-hour urine microalbuminuria assay is requested to confirm the diagnosis. The values defining microalbuminuria are as follows:
  - Urine sample: 20-200 mg/mL
  - 24-hour urine: 30-300 mg/24h (J.M. Halimil et *al* ; 2008)
- **Fasting creatinine level**. From this, it is recommended to calculate creatinine clearance, which provides information on glomerular filtration rate (Table 1) (Dr Edmond Renard; 2015).

Table1 : Interpretation of glomerular filtration rate (GFR)

GFR (ml/min)	Description
> 90	Normal or increased GFR
60 à 90	GFR slightly reduced
30 à 60	Moderate renal insufficiency
15 à 30	Severe renal failure
< 15	End-stage renal failure

- **Urea measurement** is considered an imprecise marker of renal function. A rise in plasma urea levels is therefore observed in cases of renal failure. (JC Villeneuve ; 2010

4. Risk factors for diabetic nephropathy

Not all diabetics develop diabetic nephropathy at the same rate. Genetic factors have been shown have a strong influence (genetic predisposition). In addition, the risk of disease increases with inadequate glycemic control and high blood pressure. Smokers are also greater risk than non-smokers

5. The different stages of diabetic nephropathy

Depending on the progression of nephropathy, it can be divided into five stages (Pierre Fontaine; 2014):

- **Stage I: Functional nephropathy or glomerular hypoperfiltration**

Characterized by increased kidney size, glomerular volume and glomerular filtration rate. While blood pressure and albuminuria are both normal.

- **Stage II: Silent nephropathy**

Histological renal lesions with no clinical translation. These lesions are specific to diabetes and have been described on the basis of renal biopsies. They alter the basal membrane of capillaries in the glomerulus, which progressively allow molecules of increasing size to filter through.

- **Stage III: Incipient nephropathy**

Defined by increased glomerular filtration rate, microalbuminuria (urinary albumin excretion of 30 to 300 mg/24 h) and increased blood pressure.

- **Stage IV: Established nephropathy or chronic nephropathy**

Characterized by the presence of macroalbuminuria or proteinuria (urinary albumin excretion exceeding 300 mg/24 h), histological lesions, reduced glomerular filtration rate, increasing proteinuria and arterial hypertension (> 140/90 mm Hg).

- **Stage V: End-stage renal disease (ESRD)**

Characterized by glomerular obstruction and glomerular filtration rate < 15 ml/min: requires dialysis.

6. Renal failure as a consequence of diabetic nephropathy

6.1. Definition of renal failure (RF)

Renal failure is the result of kidney disease, characterized by a reduction in the number of nephrons - functional units whose main component is the glomerulus, the small sphere where blood filtration takes place and primary urine is produced. It results in elevated blood levels of creatinine and urea.

6.2. Classification of renal failure

6.2.1. Acute renal failure (ARF)

This is a form of renal failure in which the loss of renal function is sudden but generally reversible. Depending on the mechanisms involved, there are three types of acute renal failure.

a. Functional acute renal failure

It is due to hypovolemia (reduced circulating blood volume, resulting in lower blood flow to the kidneys), and can therefore result either from a drop in blood pressure, caused by hemorrhage, heart failure or shock, or from severe dehydration linked to diarrhea or vomiting. This form of renal failure causes no damage to kidney tissue, and disappears with the disorder responsible.

b. Acute organic renal failure

It is caused by anatomical alterations to the tubules (acute tubular necrosis), interstitial tissue (acute interstitial nephritis), glomeruli (glomerulonephritis) or vessels (vascular nephropathy) of the kidney. These lesions may be due to intoxication (drugs, iodine products used for X-ray examinations), an allergic reaction, an infectious or immunological process, and so on

c. Acute obstructive renal failure

Acute obstructive renal failure is caused by an obstruction (stone, tumor) in the excretory tract (pelvis, ureters, bladder)

6.2.2. Chronic renal failure (CRF)

Defined by a prolonged, often permanent, reduction in exocrine and endocrine renal function. It is essentially expressed by a reduction in glomerular filtration rate (FG), with an increase in creatininemia and blood urea (uremia) due to reduced creatinine clearance. It can lead to end-stage renal disease (ESRD), requiring extra-renal replacement therapy (ERT) via hemodialysis or peritoneal dialysis and/or renal transplantation.

6.3. Symptoms of renal failure

The most revealing clinical sign of acute renal failure is anuria (cessation of all urine production by the kidneys). However, urine volume may only be reduced, or may even remain normal: in the latter case, we speak of acute renal failure with preserved diuresis. Some particularly serious manifestations may occur: hyperkalemia (which can lead to serious cardiac problems), acute pulmonary edema, acidosis, hyponatremia

Chronic renal failure is almost always complicated by anemia, due to reduced secretion of erythropoietin (a hormone stimulating the production of red blood cells by the bone marrow) by the kidney, leading to fatigue, shortness of breath and difficulty in exerting physical effort. Arterial hypertension is frequently observed.

6.4. Treatment and prevention of diabetic nephropathy

Glycemic regulation, treatment and control of high blood pressure are very important to avoid ND. In fact, drug therapy generally involves preparations from the ACE inhibitor group. The use of these preparations slows the progression of nephropathy.

Nowadays, it is possible to prevent the onset of kidney damage in diabetics. Optimum treatment and regular monitoring of blood pressure, blood sugar and microalbuminuria are important in this context. Patients should also avoid drugs with renal side effects. These include certain anti-inflammatory analgesics.

Materials and methods

A. Hardware

1. Study framework

This is a prospective study carried out in the medical analysis laboratory

2. Study population

Our study involved 65 diabetic patients with or without diabetic nephropathy. Patients were selected on the basis of gender, age and renal complications.

B. Methods

1. Data collection

Clinical data were collected using a file described by the physician. We measured the biochemical parameters studied only in blood, namely glycemia, HBA1c, creatinine, urea and uric acid.

2. Venous blood sampling

Blood sampling is carried out as follows: we apply a tourniquet around the forearm to make the vein protrude, then clean the skin with an alcohol-soaked cotton pad before pricking with a sterile syringe. The blood collected was either placed in dry tubes, or collected on heparin or EDTA. After detachment, the clotted blood was centrifuged at 5000rpm for 20 minutes. The serum is then collected for the various biochemical assays.

3. Methods for measuring biochemical parameters

3.1. Blood glucose test

3.1.1. Blood glucose: (fasting blood glucose)

This is an enzymatic method based on glucose oxidase and peroxidase. Results are automatically calculated by the Konelab 20 E analyzer using a calibration curve.

Principle

This method uses glucose oxidase (GOD) and a modified Trinder reaction catalyzed by peroxidase (POD). Glucose is oxidized to gluconic acid by glucose oxidase, with the formation of an equimolar amount of hydrogen peroxide. In the presence of peroxidase,

4-aminoantipyrine and phenol undergo a hydrogen peroxide oxidation reaction to give rise to a red quinoneimine dye. The intensity of the color produced by this reaction is measured at 510 nm and is proportional to the glucose concentration in the sample

$$\text{Glucose} + O_2 + H_2O \xrightarrow{\text{GOD}} \text{Acide gluconique} + H_2O_2$$

$$2\,H_2O_2 + \text{Phénol} + \text{4-Amino-antipyrine} \xrightarrow{\text{POD}} \text{Quinonéimine rouge} + 4H_2O$$

How it works

1- Place reagents at room temperature.
2- Pipette into test tubes according to Table 2 :

Table2 : Procedure for preparing solutions for blood glucose determination

	White	**Standard**	**Sample**
Glucose standard	--	10µl	--
Sample	--	--	10µl
Responsive to work	1ml	1ml	1ml

3- Shake well and incubate tubes for 10 min at room temperature (16 to 25°C) or for 5 min at 37°C.
4- Read the absorbance of standard and sample against white at 510 nm. The color is stable for at least 2 hours.

Calculation

Glucose = (Sample D.O / Standard D.O) × n

With n= 1g/l

Reference interval [0.7 - 1.1] g/l

This method requires the following reagents:

Reagent 1(buffers) : - Phosphate buffer (pH= 7.5) 100 mmol/l

- Buffer solution: Phenol 5 mmol/l

Reagent 2 (enzymes) : - Glucose oxidase > 10 U/l

- Peroxidase >1 U/l

- 4-Amino-antipyrine 0.4 mmol/l

Reagent 3 (standard) : - Glucose 1g/l

3.1.2. Glycated hemoglobin (HBA1c):

This is an automated ion exchange chromatographic technique (HPLC: HLC-723GX) (photo 1).

Principle

Based on the principle of high performance liquid chromatography or HPLC, the analyzer uses a cation exchange column to separate hemoglobin components according to differences in ionic interactions. The different components of hemoglobin are rapidly separated. A progressive gradient combined with a linear gradient with three different ionic concentrations are used for separation.

When the analysis is complete, the chromatogram and the percentages of the various hemoglobin fractions are automatically printed.

Application this chromatographic method requires the use of three elution buffers for separation and a washing solution.

Reference interval: less than 6.2%.

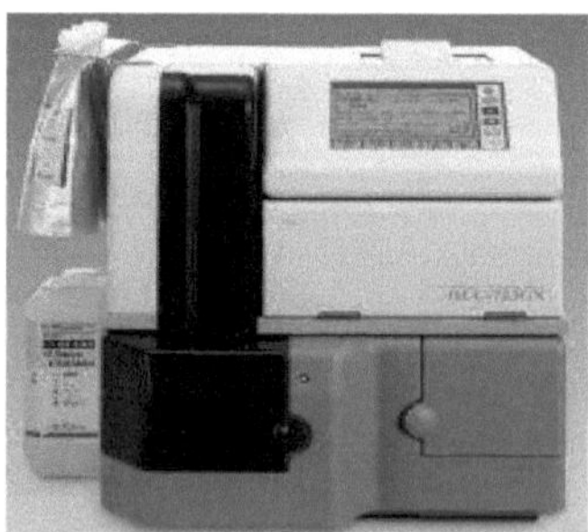

Photo1: The glycated hemoglobin meter

3.2. Renal check-up

3.2.1. Creatinine

This is an automated colorimetric method: kinetic test without deproteinization according to the Jaffé method, using a Konelab 20 E analyzer.

Principle

In the presence of picric acid and in alkaline solution, creatinine forms a red-orange colored complex. The variation in absorbance of this complex, measured at specific times, is proportional to the creatinine concentration (Jaffé M.; 1886), measured at 492 nm. Serum and plasma samples contain proteins that react non-specifically in the Jaffé method (Jaffé M.; 1886).

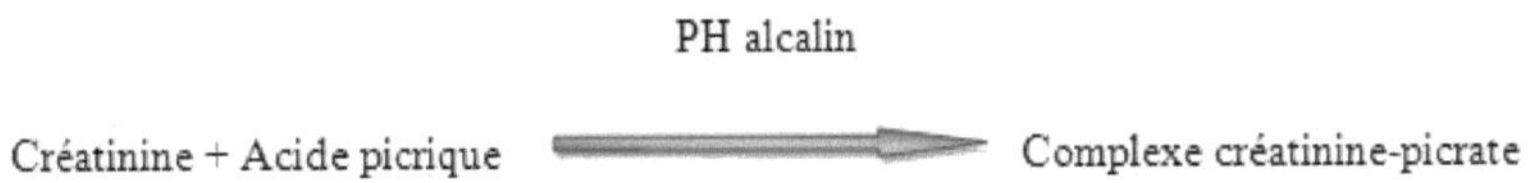

How it

Wavelength: ....492 nm (490 - 510)

Temperature:.....................................25 - 30 or 37 °C

Cell:001 cm thick

- Zero the spectrophotometer on air or distilled water.

Table3 : Procedure for preparing solutions for creatinine determination

	Standard	Sample
Standard	100µl	--
Sample	--	100µl
Responsive to work	1ml	1ml

Mix and read DO1 optical densities after 30 sec. Then read DO2 exactly 1 minute later.

Calculation

Calculate ΔDO = DO2 - DO1 for standard and samples.

Creatinine = (ΔD O Sample / Δ D O Standard) ×n

Where: n = 20 mg/l

Reference interval Man: [7-14] mg/l

Female: [6 -11] mg/l

This method requires the use of three main reagents:

Reagent 1: Sodium hydroxide 04 mol/l

Reagent 2: Picric acid 25 mmol/l

Reagent 3: standard creatinine 20 mg/l

3.2.2. Urea

This is an automated enzymatic method based on urease and glutamate (Konelab 20 E analyzer).

Principle

Urea is hydrolyzed in the presence of water and urease to give rise to ammonia and carbon dioxide. In the presence of glutamate dehydrogenase (GLDH) and NADH in reduced form, ammonia combines with α-ketoglutarate to form L-glutamate.

$$\text{Urée} + H_2O \xrightarrow{\text{Uréase}} 2\,NH_3 + CO_2$$

$$2\,NH_4^+ + \text{Oxo-2-glutarate} + NADH \xrightarrow{\text{GLDH}} \text{L-Glutamate} + NAD^+ + 2\,H_2O$$

The decrease in absorbance due to the conversion of NADH to NAD^+ , measured for a given time at 340 nm, is proportional to the urea concentration in the specimen.

How it works

Wavelength: ....340 nm

Temperature: .... 25-30-37°C

Cell:001 cm thick

- Zero the spectrophotometer on air or distilled water.

Table4 : Procedure for preparing solutions for urea determination

	Stallion	Dosage
Working solution	1ml	1ml
Preincubate at the chosen temperature (25, 30 or 37° C).		
Reagent 3 (standard)	10µl	--
Sample	--	10µl

Mix, measure OD decrease between: t = 20 seconds and t = 80 seconds.

Calculation

Calculate ΔDO = DO2 - DO1 for standard and samples.

Urea = (ΔD.O. Sample / ΔD.O. Standard) x n

With n= 0.50 g/l

Reference interval [0.15-0.45]

This method requires the use of 3 types of reagents:

Reagent 1(buffers) : - Tris buffer (pH 8.0) 100 mmol/l

Reagent 2 (enzymes) : - Urease > 140 U/ml

- GLDH >140 U/ml
- NADH 1.5 mmol/l
- Oxo-2glutarate 5.6 mmol/l

Reagent 3 (standard) : - standard urea 8.325 mmol/l

3.2.3 Uric acid

This is an automated enzymatic photometric test with "TBHBA" (2,4,6-tribromo-3-hydroxybenzoic acid) (Konelab 20E analyzer).

Principle

Uric acid is oxidized to allantoin by uricase. The hydrogen peroxide thus formed reacts with 4-aminoantipyrine and 2,4,6-tribromo-3-hydroxy-benzoic acid (TBHBA) to form quinone imine (DiaSys; Diagnostic Systems).

The resulting increase in absorbance at 520 nm (500-550 nm) is proportional to the concentration of uric acid in the sample.

$$\text{Acide urique} + 2\,H_2O + O_2 \xrightarrow{\text{Uricase}} \text{Allantoïne} + CO_2 + H_2O_2$$

$$\text{TBHBA} + \text{4-Aminoantipyrine} + 2\,H_2O_2 \xrightarrow{\text{POD}} \text{Quinone imine} + 3\,H_2O$$

How it works

Wavelength:........................ 520 nm - 550

Bowl......................................1 cm

Temperature............................... 20 - 25 °C or 37 °C

Zero the spectrophotometer on the reagent blank.

Table5 : Procedure for preparing solutions for uric acid determination

Tubes	White	Sample/standard
Sample/Standard	--	20 µL
Distilled water	20 µL	--
Reagent 1	1000 µL	1000 µL
	Mix, incubate 5 min.	
Reagent 2	250 µl	250 µl

Mix, incubate 30 min at 20 -25°C or 10 min at 37°C. Read absorbance against reagent blank within 60 min.

Calculation

Uric acid = (D.O. Sample / D.O. Standard) x n

Where: n = 60 mg/l.

Reference interval 30 - 70] mg/l

This method requires the use of three main reagents:

Reagent 1(buffers) : - Phosphate buffer (pH 7.4) 100 mmol/L

- TBHBA (2, 4, 6-tribromo-3-hydroxybenzoic acid) 1.15 mmol/L

Reagent 2 (enzymes) : - Uricase N ≥ 150 U/L

- Peroxidase N ≥ 10 kU/L

-4-Aminoantipyrine N 1.5 mmol/L

-K4 [Fe(CN)$_6$] N 50 µmol/L

Reagent 3 (standard) : - Uric acid 0.06 g/L (357 µmol/L)

4. Statistical study

Statistical analyses were carried out using Excel software (calculation of means and standard deviations). Student's t tests were used for comparisons of means. For all analyses, a p-value of less than 0.05 was considered significant.

Results and discussion

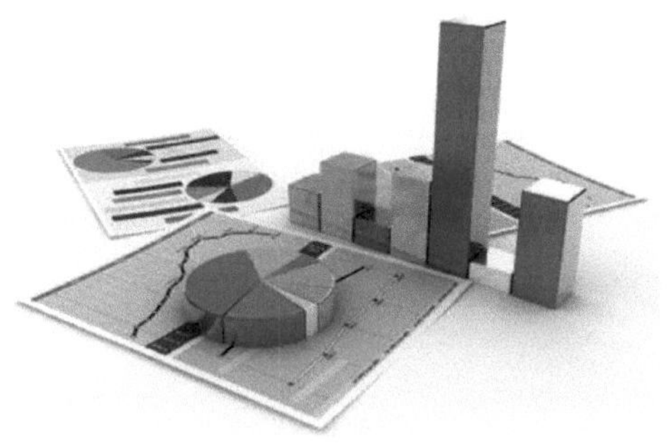

1. Epidemiological study

1.1 Patient distribution by age

A breakdown of patients by age group is shown in figure 1:

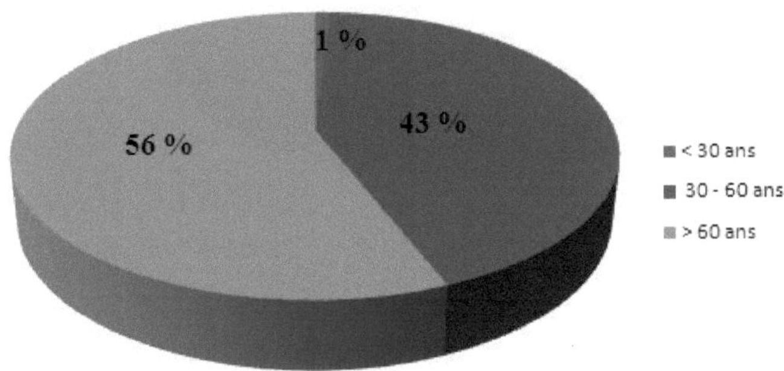

Figure1 : Age distribution of patients (n=65)

We found that 43% of patients were between 30 and 60 years of age, 56% were over 60 and only 1% were under 30. In our population, we found an average age equal to 62.78 ± 13.70. (Fig.1)

These results are similar to the study by N. Khélifi et *al*, 2011 who found a mean age equal to 56.5 ± 13.92 and the study by Eleuchi Art et *al*, 2019 who also found a mean age of 60.06 ± 13.3.

In contrast, Ach Taieb et *al*, 2018 found an average age of 34 ±14.33 with extremes ranging from 13 to 77 years. Also, H.Zouari et *al* ,2012 showed an average age of 50 ± 14.7 with extremes ranging from 18 to 77 years.

People over the age of 60 are the most affected by diabetes. This may be explained by the fact that the body's ability to produce and use insulin diminishes with age, leading to the development of diabetes in the elderly.

1.1. Patient distribution by gender

The distribution of patients by gender is shown in Figure 2.

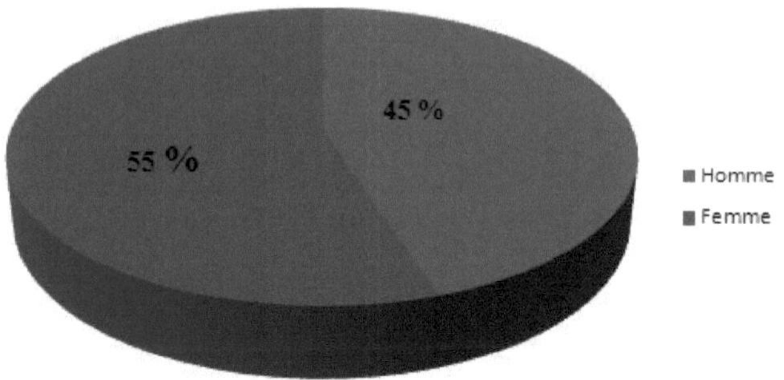

Figure2 : Distribution of patients by gender (n=65)

Our study population (n=65) was divided as follows: 29 men (45%) and 36 women (55%), with a sex ratio of 0.8. We have thus shown that the risk of diabetic disease is higher in women than in men (Fig.2).

In line with our results, a female predominance was confirmed in the study by H.Zouari et *al*, 2012 (65% women and 35% men) and E.B. Ould Isselmou et *al*, 2010 (45.2% men and 54.8% women). This was explained according to Frank Mauvais-Jarvis et *al*, 2017 by the effect of female hormones.

In fact, via their receptors, estrogens improve both insulin secretion and insulin sensitivity, while protecting pancreatic β-cells against metabolic stress. The increased risk of developing diabetes (particularly type 2), has been observed at menopause when estrogen falls.

In contrast, the study by N. Khélifi et *al*, 2011 showed that men were more affected by diabetes than women (53% vs. 47% respectively).

1.2. Distribution of patients according to renal complications

With regard to renal complications, we have shown that among the population studied, only 7 cases equivalent to 11% presented renal complications with varying degrees of severity, while the remaining 58 cases, i.e. 89%, had normal renal function (fig.3). As a result, the incidence of renal complications in diabetic subjects is 11%. This high figure shows that nephropathy is a kidney complication that can affect diabetics. Our results are similar to those of H . Ibrahim et *al*, 2011 where 21.5% of his study population had ND. In contrast, N. Khélifi et *al*, 2011 showed that 60% of his study population had ND

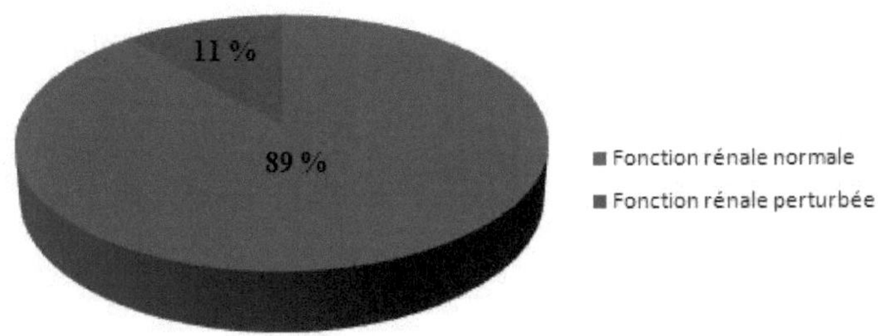

Figure3 : Distribution of patients according to renal complications (n=65)

We have also shown that patients with normal kidney function are distributed as follows: 50% male and 50% female, whereas patients with ND are all female (Fig.4).

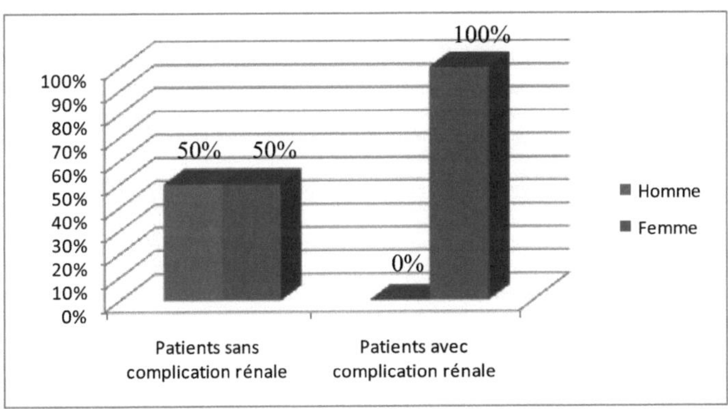

Figure4 : Gender distribution of patients with and without renal complications (n=65)

This may be due to the fact that women are more at risk of developing metabolic diseases, mainly diabetes, because of the production of sex hormones and the change in hormonal profile characteristic of the menopause.

2. Study of plasma biochemical parameters

2.1.Study of glycemic markers

2.1.1. Fasting blood glucose

Fasting blood glucose results for patients consulting our laboratory during the internship period are shown in Table 6.

Table6 : Average glycemic markers of patients studied.

	Usual values	Mean ± standard deviation
Blood glucose (g/l)	[0,7 - 1,1]	1,98 ± 0,76 (n=65)
HBA1c (%)	Less than 6.2	8,4 ± 1,70 (n= 54

We noted an increase in fasting glycemia in patients (n=64) with an average equal to 1.98 ± 0.76 g/l. (Table 6)

What's more, 56 cases (86%) had above-normal fasting blood glucose levels, compared with only 9 cases (14%). (Fig.5)

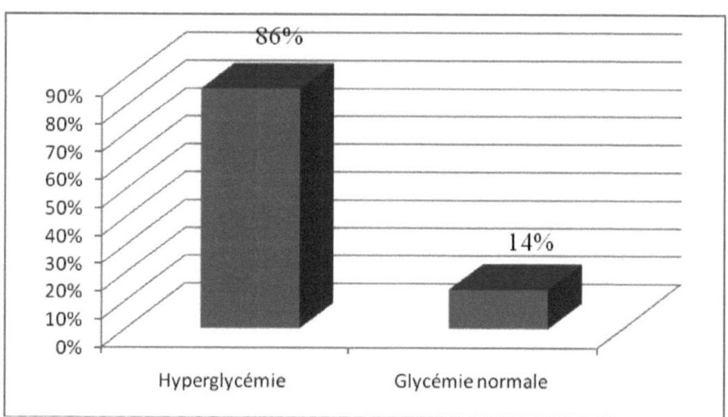

Figure5 : Distribution of patients by blood glucose level (n=65)

According to Ach Taieb et *al*, 2018; the average blood glucose level for his study population is equal to 3.09 ± 1.01 g/l.

Our results also show that blood glucose levels are higher in patients with diabetic nephropathy (2.52±0.95 g/l) than in patients without diabetic nephropathy (1.92±0.72 g/l). (Table 7)

Table7 : Glycemic markers in patients with and without renal complications

	Subjects without renal complications	Subjects with renal complications
Blood glucose (g/l) (n=65)	1,92 ± 0,72 (n=58)	2,52 ± 0,95 (n=7)
HBA1c (%) (n=54)	8,28 ± 1,60 (n=47)	9,2 ± 2,24 (n=7)

This is confirmed by Roussel, 2011, who has shown that hyperglycemia plays a causal role in the pathophysiology of the initial stages of diabetic nephropathy and worsens renal damage. In fact, high blood sugar or glucose levels can damage artery membranes, leading to a rise in blood pressure. This in turn forces the kidneys to filter an excessive quantity of blood, leading to nephron exhaustion and degradation, resulting in ND

2.1.2. Glycated haemoglobin

The average glycated hemoglobin levels of 54 of the diabetic patients studied are shown in Table 6.

It is equivalent to 8.4 ± 1.70%, which is well above the norm (<6.2%) (Table 6). Moreover, only 15% of our population showed normal HBA1c levels, while 85% had HBA1c values above the threshold (Fig.6).

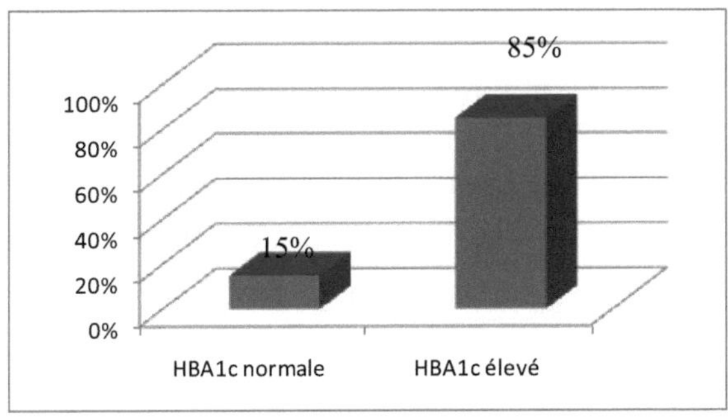

Figure6 : Distribution of patients by HbA1c level (n=54)

These results are in line with those found by Ibrahim Hamat et *al*, (2016) where 74.5% of his population had glycated hemoglobin exceeding the value assumed to be normal. While Sylviane Deletre et *al*, (2016) found a mean HbA1c for all patients equal to 6.85 ± 1.26%.

In contrast, Ach Taieb et *al*, 2018 showed a mean HBA1c equal to 12.17± 2.31%.

However, HbA1c control is closely linked to the prevention of diabetes-related micro and macrovascular complications, particularly ND. (HAS, 2013)

We have shown that the mean HBA1c is higher in those with diabetic nephropathy (9.2±2.24%) than in patients without diabetic nephropathy (8.28±1.60%). (Table 7)

We suggest that a relationship may exist between increased glycated hemoglobin levels and the development of diabetic nephropathy.

2.2. Study of renal markers

2.2.1 Creatinine

Renal markers (creatinine, urea and uric acid) in diabetic patients are shown in Table 8. We found that the mean plasma creatinine level was 16.90 ± 15.59 mg/l, well above the norm (Table 8).

Table8 : Average creatinine, urea and uric acid levels of patients studied.

	Usual values	Mean ± standard deviation
Creatinine (mg/l)	[7 - 14] : Male [6 - 11] : Female	16,90 ± 15,59 (n=59)
Urea (g/l)	[0,15 - 0,45]	0,71 ± 0,47 (n=27
Uric acid (mg/l)	[30 - 70]	71,16 ± 25,53 (n=31)

Indeed, 27 patients in our population, or 46%, had normal creatinine levels, while 47% had hypercreatinemia with a maximum value of 86 mg/l, and 7% had hypocreatinemia. (Fig.7)

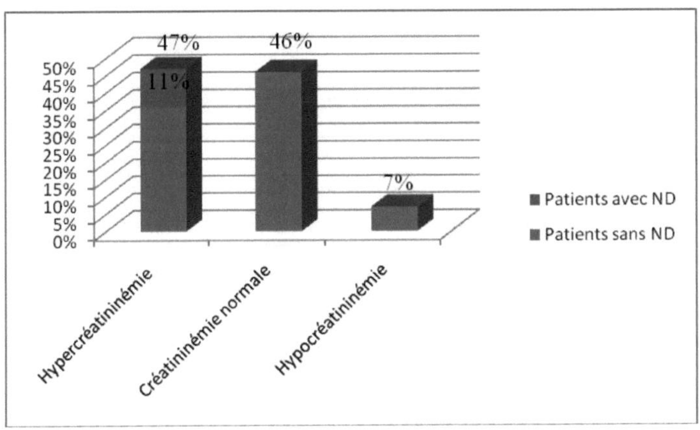

Figure7 : Distribution of patients with and without ND according to creatinine level (n=59)

Creatinine has long been considered the best endogenous marker of glomerular filtration.

According to our results, 47% of our subjects had supraphysiological creatinine levels, of whom only 11% had diabetic nephropathy (fig.7). This means that creatinine is an excellent marker for assessing renal function. A correlation could be established between plasma creatinine levels and the degree of renal complication. On the other hand, several studies have shown that creatinine cannot be used to diagnose early renal failure (Dussol, 2011), particularly in elderly subjects, as its value depends on the subject's sex, muscle mass and diet (Roland et *al*, 2011).

2.2.2 Urea

Based on Table 8, we have shown that the mean uremia measurement in 27 patients only is low, equal to 0.71 ± 0.47 g/l.

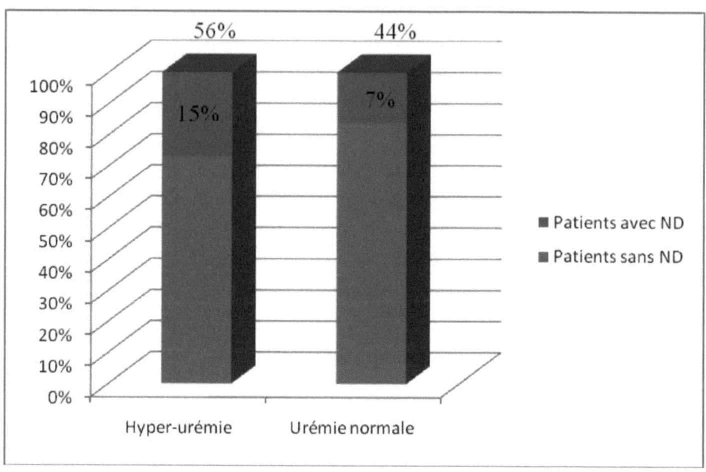

Figure8 : Distribution of patients with and without ND according to urea levels (n=27)

In fact, we found that 44% of them had normal physiological levels, while 56% had hyper-uremia with a maximum value equal to 1.74 g/l. (fig.8).

Our uremia results show that almost half of our population have hyper-uremia. This means that their kidney functions are impaired, but this impairment may be diabetic or of other origins. It has been suggested that urea levels may increase in proportion to the degree of kidney damage.

In addition, we found that 15% of hyper-uremic patients have ND (fig.8). Jamoussi et *al*, 2005 have shown that any increase in blood urea reflects a deficit in the excretory function of the kidneys. The more renal function is impaired, the more urea accumulates in the blood and becomes a toxic factor.

2.2.3 Uric acid

Table 8 shows that the mean uric acid level in 31 patients was 71.16 ± 25.53 mg/l. Thus, 45% of subjects have hyperuricemia (>70 mg/l) with a maximum value of 123 mg/l and 55% have normal uricemia. (Fig.9)

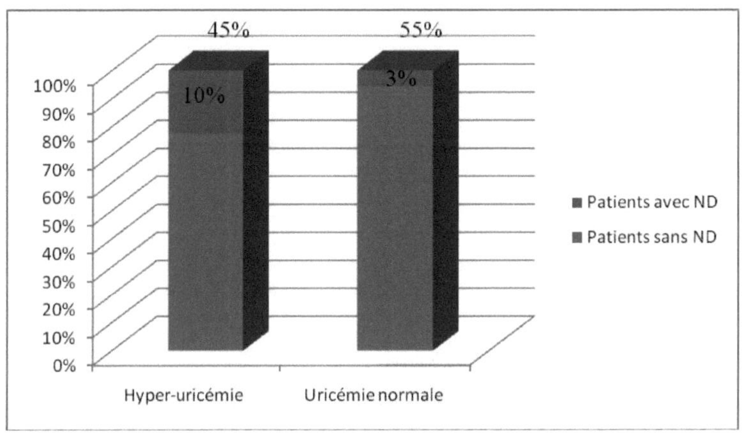

Figure9 : Distribution of patients with and without ND according to uric acid level (n= 31)

These results are close to those found by Bouattar et *al*, 2009 where they showed a mean level equal to 47±12.3 mg/l in the uncomplicated diabetic group, and 76±24 mg/l in renal failure cases.

Increasing uricemia is explained by the linear progression of impaired renal function and the inability to eliminate catabolic waste products.

We have shown that 10% of patients with hyperuricemia are ND. (Fig.9)

According to Kang et *al*, 2002, hyperuricemia is considered a marker of renal dysfunction rather than a risk factor for progression of renal damage. Similarly, the 2012 study by N. Khélifi and colleagues showed that uric acid correlates with renal function in diabetics.

Conclusions and outlook

Conclusions and outlook

The present work is a prospective study of diabetic patients who consulted the medical analysis laboratory with the aim of studying the clinical and biochemical parameters associated with diabetes-related complications, in particular renal impairment by diabetic nephropathy.

It's clear that, far from the histological changes that initiate kidney damage remain asymptomatic for several years, late diagnosis, either biologically by biochemical analysis or clinically by symptoms resulting from more advanced complications, remains a problem for effective management. Diagnosis of the disease through analysis and monitoring of parameters remains a top priority, especially in early stages of renal involvement.

The results obtained through our study showed that :

- 43% of patients were aged between 30 and 60, 56% were over 60 and only 1% under 30.
- the risk of diabetic disease is higher in women than in men (45% men vs. 55% women).
- 11% had renal complications and were 100% female. While 89% had normal renal function, divided as follows: 50% men and 50% women.
- 86% had fasting blood glucose levels above normal, while only 14% had normal blood glucose levels. The mean blood glucose value was 1.98 ± 0.76 g/l.
- Blood glucose levels were higher in patients with diabetic nephropathy (2.52±0.95 g/l) than in those without (1.92±0.72 g/l).
- 15% showed normal HBA1c levels, while 85% had HBA1c values above the threshold. The mean HBA1c was 8.4 ± 1.70%.
- Mean HBA1c was higher in those with ND (9.2±2.24%) than in patients without ND (8.28±1.60%).
- 46% had normal creatinine levels, 47% had hypercreatininemia, of which 11% had ND and 7% had hypocreatinemia. Mean creatinine was 16.90 ± 15.59 mg/l.

- 44% had normal urea levels, while 56% had hyper-ureaemia, of which 15% had ND. The mean urea value was 0.71±0.47 g/l.
- 55% had normal uricemia and 45% had hyperuricemia, including 10% with ND. Mean uricemia was 71.16 ± 25.53 mg/l.

This study remains preliminary, and requires further in-depth study. In this context, and as perspectives for our work, it would be interesting to :

- Continue research by undertaking work on a larger population.
- Evaluate other parameters in the glycemic and renal profile, such as microalbuminuria, proteinuria, urine chemistry, etc.
- Continue research on an animal model to better understand the mechanisms of action of these events that lead to the development of diabetic nephropathy.
- Move on to more sophisticated techniques than biochemistry, such as radioisotope methods and DMSA (dimercaptosuccinic acid) scintigraphy.

References

References

A

- Association suisse de diabète ; www. diabetesuisse.ch .
- Abid Mohamed, président de la société tunisienne d'endocrinologie, diabète et malades métaboliques (STEDIAM) Par African Manager, 9 novembre 2017.
- Agence nationale d'accréditation et d'évaluation en santé : SUIVI DU PATIENT DIABETIQUE DE TYPE 2 A L'EXCLUSION DU SUIVI DES COMPLICATIONS
  TEXTE DES RECOMMANDATIONS ; Janvier 1999.
- Ach Taieb1, &, Asma Ben Cheikh2, Yosra Hasni1, Amel Maaroufi1, Maha Kacem1, Molka Chaieb1, Koussay Ach1 (2018) : Etude sur le diabète aigu cétosique inaugural dans un hôpital du Centre Est Tunisien ; Service d'Endocrinologie et Diabétologie CHU Farhat Hached, Sousse, Tunisie, 2Service d'Hygiène Hospitalière CHU Farhat Hached, Sousse, Tunisie.
- Agence de la santé publique du Canada - Diabetes in Canada: Facts and figures from a public health perspective ; 2011.

B

- BioSystems S.A. Costa Brava 30,08030 Barcelona (Spain).
- Biolabo ; fabricant Biolabo SAS, les hautes Rives 021601, Marzy, France. www.biolabo.fr
- Bouattar T., Ahid S., Benasila S., Mattous M., RhooH. , et al. (2009) : .Les facteurs de progression de la néphropathie diabétique : prise en charge et évolution. Néphropathie et Thérapeutique. 5 :181-87

C

- Catalogue en PDF – Biomaghreb ; www.biomaghreb.com

D

- Dr Karim Gariani, Dr Christel Tran, Pr Jacques Philippe : Hémoglobine glyquée nouvel outil de dépistage ? Rev Med Suisse 2011 ; 7 : 1238-42.
- Docteur Jocelyne MAURIZI-BALZAN, Professeur Philippe ZAOUI-Mars 2004

Dr Edmond Renard Médecin biologiste, DFG (DÉBIT DE FILTRATION GLOMÉRULAIRE) Mai 2015.

- DiaSys Diagnostic Systems GmbH ; IVD Alte Strasse 9 65558 Holzheim (Allemagne)
- Dr Frank Mauvais-Jarvis ; Diabète : traiter les hommes et les femmes différemment?, 5 décembre 2017.
- Dussol B. (2011) : Méthodes d'exploration de la fonction rénale : intérêt et limites des formules permettant d'estimer la fonction rénale ; Immuno-analyse et biologie spécialisée. 26 : 6-12 .

E

- Équipe de professionnels de la santé de Diabète Québec, Mai 2014 (mise à jour novembre 2019).
- Eleuchi A, Bachrouch M, B Hassine A, B Mefteh N, Bayar M, Naffeti E, Ammar Y. Service des Urgences-SMUR CHU Maamouri Nabeul, Tunisie. **INFECTION URINAIRE CHEZ LE DIABÉTIQUE : ÉTUDE DE 46 CAS -P214**
- E.B. Ould Isselmou, N. Abdoul, L. Abdoulaye, P. Abdoulaye, K. Mourtalla, D. Boubacar, T. Moreira Diop (2010) ; Aspect épidémiologique, et diagnostique de la néphropathie chez le diabétique : étude rétrospective - P35.

H

- Haute autorité de santé, Prévention et dépistage du diabète de type 2 et des maladies liées au diabète, Octobre 2014
- H. Zouari, I. Farhat, J. Abdelhedi, R. Jomli, R. Damak, F. Nasef, N.Zouari (2012) : Trouble boulimique, trouble anxiodépprisif et diabète : à travers une étude de 65 patients diabétiques.- P341
- H. Ibrahim, F. Ben Mami, H. Jelassi, A. Ben Hammouda, A.Temessek, S. Dakhli, A. Achour (2011) : Profil des patients diabétique au cours d'un mois d'urgence.
- Haute autorité de santé, Stratégie médicamenteuse du contrôle glycémique du diabète de type 2, Janvier 2013.

J

- J.M. Halimi, S. Hadjadj, V. Aboyans, F.A. Allaert, J.Y. Artigou, M. Beaufils, G. Berrut, J.P. Fauvel H. Gin, A. Nitenberg, J.C. Renversez, E. Rusch, P. Valensi, D. Cordonnier (2008) : Microalbuminuria and urinary albumin excrétion: French guidelines.
- Jaffé M. (1886).Uber den Niederschlag. WelchenPikrinsaure in normalern. Harncrzeugtun du bereineneneReaction des Kreatinins Z. PhysiolChem, Vol. 10, P. 391-400.
- Jamoussi K., Ayedi F., AbidaN., Kamoun K., et al. (2005) : Profil lipidique dans l'insuffisance rénale chronique au state d'hémodialyse. Pathologie Biologie. 53 : 217-20.
- JC Villeneuve, 2010, Physiologie et physiopathologie rénales.

K

- Kang D.H., Nakagawa T., Feng L., Watanabe S., Lin Han et al. (2002): A role for uric acid in the progression of renal disease. J.Am. Soc. Nephrol. 13 : 2888-97.

L

- La revue d'information de la Société Francophone du Diabète Paramédical Les stades de la néphropathie diabétique par Pierre Fontaine CHRU Lille, 2014
- Larousse Médical encyclopédie.

N

- Nam H.-C. et al (2013). Atlas du diabète de la FID, 6ème édition, Fédération internationale du diabète, P. 22-110.

- Nam H.-C. et al (2017). Atlas du diabète de la FID, 8ème édition, Fédération internationale du diabète.

- Navin Jaipaul, janv. 2018 MD, MHS, Loma Linda University School of Medicine ; Le manuel MSD Version pour professionnel de la santé.

- N. Khélifi, E. Khadhraoui, N. Trabelsi, A. Falfoul, F. Ben Mami, S. Dakhli, A. Achour (2011) ; La néphropathie diabétique : profils clinique et épidémiologique - P232, Service (C) de diabétologie et nutrition à l'Institut National De Nutrition, Tunis, Tunisie.

O

- **Organisation mondiale de la santé 2016** (World Heath Organization 2016): **Rapport mondial sur le diabète situation générale.** www.who.int

R

- Roussel R. (2011) : Histoire naturelle de la néphropathie diabétique. Médecine des maladies métaboliques Vol. 05. Suppl.1 :8-13.
- Roland M., Guiard E., Kerras A., Jacquot C. (2011) : Pourquoi la clairance da la créatinine doit-elle céder la place aux formules d'estimation du DFG ? ; Revue francophone des laboratoires. 429 Bis : 28-31.

S

- Salma-Chaudhry A., Mavromati M., Golay A. (2013). Diabète type 2, Hôpitaux universitaires de Genève, P. 2.
- Stengel B., Billon S., Dijk PC., Jager KJ., (2003). Trends in the incidence of renal replacement therapy for end-stage renal disease in Europe. 1990-1999. Nephrol. Dial. Transplant. 18 (9): 1824-33.
- Sylviane Dletre, Martal Coutaz ; Rev Med Suisse 2016, volume 12,461-466.

T

- Thermo Fisher Scientific.
- Tosoh Automated Glycohemoglobin Analyzer HLC-723G8 Guide de l'opérateur (Mode d'analyse ß-Thalassemia).

Renal complications in diabetics

Summary:

The aim of our work was to determine the distribution of diabetes according to sex, age and renal complications, and to study certain biochemical parameters of the glycemic and renal balance of people with diabetes and/or diabetic nephropathy (n=65). We found that the risk of diabetic nephropathy was higher in women than in men, and that 11% developed renal complications. 86% had above-normal fasting blood glucose levels, and blood glucose levels were higher in those with ND (2.52±0.95 g/l) than in patients without ND (1.92±0.72 g/l). 85% of patients had HBA1c values above threshold, and mean HBA1c was higher in those with ND (9.2±2.24%) than in those without ND (8.28±1.60%). In addition, 47% had hypercreatininemia, of whom 11% had ND and 7% had hypocreatinemia. Hyper-uremia was present in 56% of patients, of whom 15% had ND, and 45% of patients had hyper-uricemia, of whom 10% had ND. Our results also show that the renal work-up is effective in estimating function and the degree of renal complication.

Key words: Diabetes, nephropathy, glycemia, HBA1c, creatinemia, uremia, uricemia

Printed by Books on Demand GmbH, Norderstedt / Germany